Fluid Power
Educational
Series

Hydraulic Fluids

Joji Parambath

Hydraulic Fluids

Copyright © 2020 Joji Parambath

All rights reserved

ISBN: 9798653627071

https://jojibooks.com

Disclaimer of Liability

The contents of this book have been checked for accuracy. Since deviations cannot be precluded entirely, we cannot guarantee full agreement. Only qualified personnel should be allowed to install and work hydraulic equipment. Qualified persons are defined as persons who are authorised to commission, to ground, and tag circuits, equipment, and systems following established safety practices and standards.

Table of Contents

Preface

Hydraulics is a branch of engineering science that uses the incompressible fluid medium. Significant technological advances have been made in the development of many types of fluids for meeting the exacting requirements of modern hydraulic applications. All hydraulic systems, however, have a common need for protection against harmful contaminants. Reasonable contamination control means cost-effective filtration and fluid analysis.

The textbook entitled 'Hydraulic Fluids' explains, in detail, the functions, types, characteristics, and selection of hydraulic fluids. The subsequent sections present topics on fluid contamination, the effect of contamination on fluids, fluid analysis, the fluid quality standards, and the maintenance aspects of fluids.

Many other fluid power topics are given in other textbooks under the fluid power educational series by the same author. A list of all the books is given at the end of the book. Also, please see the details at: https://jojibooks.com.

Enjoy reading the book.
Your feedback is most welcome.

JOJI Parambath

Chapter 1 | Hydraulic Fluids - Functions and Preparation

Modern hydraulic applications demand compact machines designed with tighter tolerances and that run at faster cycle times. They operate at higher pressures, temperatures, and speeds. They are designed to work with small amounts of fluids. Under these circumstances, the fluid used in a hydraulic system is subjected to severe stresses.

Functions of Hydraulic Fluids

The fluid in a hydraulic system has to carry out many essential functions. The essential functions of the fluid are to:

- transmit power,
- provide lubrication to the internal moving parts in the system,
- provide sealing between clearances between moving parts, and
- assist in the removal of contaminants and heat from the system.

Preparation of Hydraulic Fluids

Hydraulic fluid is prepared from a base stock and additives.

The base stock possesses all the essential characteristics to perform well in a particular class of hydraulic systems. Some examples of the base stock are petroleum fluids, high-water-based fluids, synthetic fluids, and vegetable fluids.

Many types of hydraulic fluid can be formulated by adding the base stock with varieties of additives to meet the exacting requirements of complex systems. Blending the base fluid with suitable additives can improve the physical and chemical properties of the fluid, and make the properties more stable even in the presence of heat, oxygen, and water. See Table A2.1 for more details on hydraulic fluid additives.

Chapter 2 | Characteristics of Hydraulic Fluids

Appendix 1 gives the basics of viscosity. Furthermore, many of the essential properties of hydraulic fluids are briefly explained in the following sections.

Viscosity

If the fluid medium in a hydraulic system is exposed to cold temperature, then it is going to be thicker, and its viscosity tends to be high. More energy is required to pump the thick fluid. Consequently, the thick fluid produces a higher pressure drop and generates excessive heat. It may lead to the sluggish operation, higher power consumption, and lower mechanical efficiency of the system. It may also produce cavitation and damage filters in the system.

If the fluid is exposed to hot temperature, then it is going to be thinner, and its viscosity tends to be lesser. The fluid that is too thin tends to rupture the fluid film between sliding surfaces in the system, produces leakages and a higher rate of friction. It also produces a higher rate of oxidation and causes a reduction in its service life.

The fluid used in a system must be thin enough to make it flow smoothly but thick enough to maintain sufficient lubricating film between sliding surfaces in the system components and to provide proper sealing. For most hydraulic applications, the hydraulic fluid must be kept at a reasonable thickness at the usual operating temperature.

Optimum Viscosity Range

Within the modern industrial systems, there is a broad range of hydraulic applications, which have varying requirements in respect of their viscometric characteristics. Figure 2.1 shows the behaviour of volumetric and mechanical efficiencies of a hydraulic system against the variations in the viscosity of the system fluid. For a typical application, the viscosity of its fluid medium should ideally remain between 16 to 36 cSt at the operating temperature, as it is in this viscosity range that the components in the system, such as pumps, valves, etc., maintain their excellent volumetric characteristics. The general rule of thumb is that the viscosity should never fall below 10 cSt [60 SSU] and rise above 400 cSt [1950 SSU], under any circumstance.

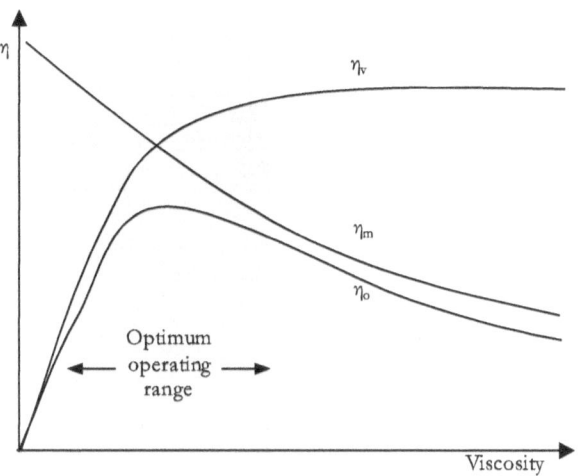

Figure 2.1 | Optimum viscosity range

Where,

η	– Efficiency
η_v	– Volumetric efficiency
η_m	– Mechanical efficiency
η_o	– Overall efficiency

Effects of Viscosity Variations

The graphical representation of Figure 2.2 highlights the significant problems associated with the use of hydraulic fluids with viscosities above and below the optimum range.

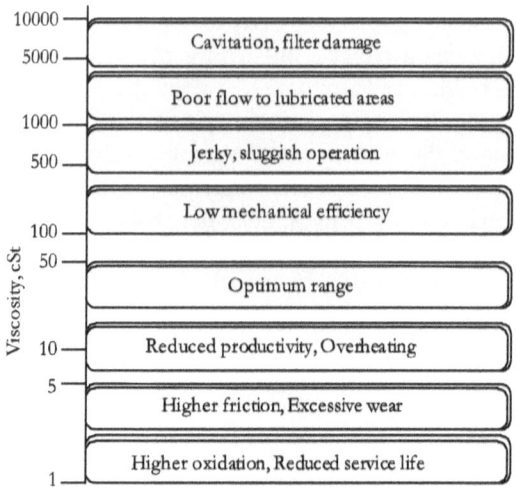

Figure 2.2 | The impacts of viscosity variations on hydraulic systems/fluids

Viscosity Index

Certain hydraulic systems are subjected to wide variations in temperatures. Mobile hydraulic systems are exposed to the outside environment. A high precision hydraulic system is also sensitive to changes in the viscosity of its fluid medium at low temperatures. A hydraulic system which is exposed to wide variations in temperature requires a fluid with high viscosity index (VI) to maintain its viscosity variations to a minimum, irrespective of changes in the temperature. Some fluids have large VIs, to begin with, and some other fluids have their VIs reinforced through VI improvers. The latter type of fluids is sometimes called as 'multi-viscosity' fluids or 'multi-grades'.

Fluid Compressibility

A good hydraulic fluid should have very low compressibility (that is, high bulk modulus) so that it remains stiff, and that helps to get a fast response from the associated system. However, the compressibility of the fluid increases as the temperature and pressure, to which it is subjected, increase. A typical mineral-based fluid undergoes about 0.5% reduction in volume for every 70 bar [1000 psi] of pressure exerted upon, up to the pressure of 300 bar [4350 psi]. Water-based fluids and synthetic fluids have higher bulk modulus as compared to that of the mineral-based fluids.

The effect of compressibility shows up as a loss of fluid volume. This volume loss represents a loss of power, as no downstream actuator is capable of recapturing the compressive energy. Therefore, fluids with low compressibility must be selected for specialised systems such as those in precision machines or servo systems.

Lubricity

A fluid provides a load-carrying film in the clearance between two relatively moving surfaces in a hydraulic component. The film prevents metal to metal contact and thus minimises friction. Under modest load conditions, petroleum fluids can satisfy the lubrication requirements of the system.

Wear Resistance

A fluid intended to be used under modest load conditions should be formulated with anti-wear additives to improve its wear resistance. An anti-wear additive is the stabilised zinc dithiophosphate (ZDP). The ZDP under highly stressed condition may produce undesirable ash. Fluid manufacturers look for environmentally-safe ashless additives to zinc-based additives.

With high loads, it is hard to maintain a sufficiently thick fluid film between the clearances of moving parts in a component of a hydraulic system. The fluid used for this type of application should be formulated with Extreme Pressure (EP) additive to improve its load-carrying property.

Oxidation Resistance

Over a period, fluid passes through various components and naturally oxidises and forms reaction products, such as acids, sludge, varnish, and gum. The exposure of the fluid to heat, water, air, and metal catalysts accelerates the natural process of oxidation. The signs of the oxidation process appear as changes in its colour, odour, and acidity level.

A superior hydraulic fluid should resist any reaction with oxygen. Better oxidation resistance can be achieved by selecting a base fluid having excellent chemical stability. Additionally, antioxidants can be used for the excellent oxidation resistance and the effective neutralisation of acids.

Corrosion Resistance

Corrosion occurs due to the reaction of moisture and oxygen in the fluid with metal surfaces. It leads to abrasive wear of the parts and increases the leakage by opening up tolerances of close-fitting parts. System rusting occurs when moisture and oxygen attack ferrous parts. A suitable rust inhibitor added to the fluid in a system protects the fluid against system rusting and chemical corrosion.

Foam Resistance

The form is a mass of air bubbles that collect at the air-liquid surface in the reservoir. The foam may be generated by the excessive churning or impingement of return fluid at the bottom part of the reservoir. If the process of foaming is excessive, the foam is likely to be drawn again into the fluid. Hydraulic fluid should, therefore, have the property of low foaming.

Stability

A fluid should have excellent thermal, chemical, and hydrolytic stabilities. Thermal Stability refers to the ability of the fluid to resist degradation when subjected to high temperatures and extreme shear. Chemical Stability refers to the ability to resist its degradation when subjected to increased chemical activities. Hydrolytic stability refers to the ability to resist its chemical decomposition in the presence of water.

Fire Resistance

The basic parameters of a fire-resistant fluid are its resistance to ignition and resistance to the propagation of the flame from the source of ignition. Flash point and fire point are the two fluid parameters used in this context.

Flash Point

Flash Point refers to the lowest temperature at which a fluid gives off enough vapours to form an ignitable mixture. This mixture is likely to generate flashes when brought into contact with a heated matter.

Fire Point

Fire Point refers to the lowest temperature at which a fluid gives off an adequate amount of vapours to its surrounding air. This amount of vapours is capable of supporting continuous combustion after the ignition.

Pour Point

Pour point refers to the lowest temperature at which a fluid can flow when cooled under the specified test conditions.

Chapter 3 | Categories of Hydraulic Fluids

As modern hydraulic systems require high-performance fluids to meet the stringent requirements of the systems, manufacturers prepare varieties of hydraulic fluids.

Petroleum-based Fluids
Petroleum (mineral-based) fluids have been the preferred energy transfer media for hydraulic systems for many years. They have excellent lubricating and corrosion-inhibiting properties, and they are low-cost fluids. Further, they are available in a broad range of viscosities. However, they are flammable. They are also toxic and not very much bio-degradable. They must be compatible with the materials of construction, such as seals. See Table A3.1(a) for more details on mineral-based hydraulic fluids.

Fire-resistant Fluids
Two basic types of fire-resistant fluid are: (1) High-water-based-fluids (HWBF) and (2) Synthetic fluids. HWBFs are very much fire-resistant due to their high water content. Synthetic fluids have an exceptional fire-resistant property, but they are costly. ISO 6743 divides fire-resistant fluids into HFA, HFB, HFC, and HFD.

- **HFA:** Fluid in water emulsions with a combustible proportion of 20% maximum
- **HFB:** Water in fluid emulsions with a combustible proportion of 60% maximum
- **HFC:** Water glycol solutions with a water proportion of at least 35%
- **HFD:** Water-free fluid on a synthetic base

Synthetic Fluids (HFD type)
Synthetic fluids are prepared from alkaline compounds that are blended with additives. Synthetic fire-resistant fluids are: (1) Phosphate esters, (2) Polyol esters, and (3) Halogenated hydrocarbons. They have good fire resistance and excellent lubrication characteristics. However, they are expensive and

often not compatible with many types of seal materials. They may also give off toxic vapours. Further, they require special disposal plan. Fire-resistant fluids are required for high-temperature, hazardous hydraulic applications.

Biodegradable Fluids
For ecologically-sensitive applications, where fluid leakage could harm the environment, there is a demand for ecologically-safe fluids. The best choice of fluids for such applications is the biodegradable type of fluids.

The criteria for defining a bio-degradable fluid is as follows. On the occurrence of the fluid spillage, 60% of the fluid breaks into harmless products. This breakage is a result of the reaction of the fluid with naturally occurring bacteria, when exposed to the atmosphere for 28 days, in a standard test. The most important base fluids of biodegradable hydraulic fluid are: (1) Synthetic esters and (2) Vegetable fluids. See Table A3.1(b) for more details on bio-degradable hydraulic fluids.

Food-grade Fluids
Devices used for manufacturing and packaging of beverages, food, cosmetics, and medicines must be hygiene specific. These devices must be designed with food-grade fluids to protect from the risk of incidental contacts of products with the fluids. Fluid manufacturers prepare the food-grade fluid from highly refined, non-toxic, Polyalphaolefin (PAO) synthetic base fluid, a highly specialised non-toxic, food-grade additive package, and a food-grade anti-microbicide. They need to be maintained to a high level of cleanliness, sanitation, and quality.

See Table A4.1 for more details on fluid classification according to DIN 51524.

Chapter 4 | Hydraulic Fluid Contamination

Hydraulic fluids are subjected to various types of contamination. Contaminants can affect the physical and chemical properties of fluids. If the contaminants are not monitored or controlled, the fluid is likely to be subjected to various kinds of failures. These failures include the deterioration of the fluid properties and consequent fluid breakdown. At the system level, the contamination can affect the performance and service life of the components and cause their erratic operation, increased heat generation, frequent fluid replacement, system failure, and higher costs.

Three critical issues concerned with the proper maintenance of hydraulic fluids are: (1) Knowing the type of contaminants, (2) means for controlling contamination, and (3) assessing the health of the fluids. Analysis of the fluid in a system can help detect an emerging problem in the system.

Contaminations in Hydraulic Fluids

Contaminants are the natural enemy of hydraulic components and systems. About 70 to 80% of the system failures are due to the adverse effects of contaminants. Even minute particles can damage components due to the existence of minuscule clearances in today's hydraulic components.

Solid Particles

It includes hard particles, such as dust, dirt, silica, and wear metals, and soft particles, such as elastomers and fibres. Silts are particles less than 5 microns. Chips are particles greater than 5 microns. Silt particles of a size corresponding to the typical clearance of components are most dangerous to a hydraulic system than larger chip particles. The system itself can generate metal particles from component wear. Abrasive particles scrap metal from the surfaces of components. The freely-circulating particles can cause premature wear of parts.

Water

Water is introduced into the system fluid by the condensation of humid air. Unprotected reservoir openings, leaking seals, and ineffective heat exchangers are other means of entry of water in hydraulic systems. Fluid can dissolve water up to its saturation point. Above the saturation point, water remains in the free or emulsified state. A mineral fluid can permit water content up to 100 ppm, that is, up to 0.01%. The moisture is capable of providing oxygen for chemical reactions.

Chemical Contaminants

They are formed by the breakdown of additives as a result of chemical reactions. The reaction products can generate other contaminants in the form of acids or oxidants in the presence of water and heat. They can cause physical and chemical changes in the additive elements. These changes can lead to the deterioration of additives and subsequent fluid breakdown.

Air

Air can enter into the fluid medium through system leaks, pump aeration, or excessive fluid turbulence in the reservoir. Next, air can exist either in the 'free state' or in the entrained state.

An air pocket trapped in a part of the system is an example of free air. Air bubbles less than one millimetre in diameter dispersed in the fluid medium is the entrained air. The entrained air can cause cavitation and foaming, as it cycles through the system. It also tends to make the system operation spongy and system response weak.

Heat

A hydraulic fluid medium is liable to deteriorate at higher temperatures and as a consequence might lose its lubricating property. Therefore, it is required to maintain the temperature of the fluid within satisfactory limits. Typically, the maximum temperature specified for fluids used in the general type of hydraulic systems is 80^0C [175^0F].

Sources of Fluid Contamination

Table 4.1 gives the sources of fluid contamination in hydraulic systems.

Table 4.1 | Sources of fluid contamination

Type	Contaminants	Example	Source
Solids	Abrasive solids	Copper, iron, silica, plastic	From wear out parts
	Non-abrasive solids	Carbon, dust, soot, cellulose, fibre, rubber, cotton, jute, nylon,	From polluted surroundings
Liquids	Dissolved water	Invisible in nature	Humid air intake
	Emulsified water	Oil turns milky white	Leakage in the heat exchanger
	Free water	Water layer separates from the fluid	Along with fresh fluid Worn cylinder seals/wipers
Gases	Free air	Oxygen, hydrogen, carbon-di-oxide, sulphur dioxide, water vapour	In-take air through breathers From pumping fresh fluid

Chapter 5 | Contamination Control

If a hydraulic system is not monitored correctly, the contamination level in the system tends to go up soon, and the system will begin to degrade. Some symptoms, such as the pump failure, premature wear, leakage, loss of fluid, sticking valves, erratic operation, solenoid burnout, chattering noise, and/or cylinder scoring, provide an early warning of the presence of excess contamination in the system.

The most favourable location for monitoring and carrying out focused contamination control activities is the system reservoir. It is the place where the system fluid gets enough dwell time, and the particles in the system tend to settle. The contamination control primarily involves the removal of particles, water, air, sludge, acid, and chemicals, from the fluid.

The particles in the fluid medium of a hydraulic system can be removed by installing correctly-sized filters at appropriate locations in the system. Magnets installed at appropriate positions in the fluid tank can remove the ferrous particles and rust matters from the fluid.

The removal of acids, sludge, gums, varnishes, and other oxidation products requires the use of an adsorbent filter with active type clay, charcoal, or activated alumina.

The problems due to the air contamination in hydraulic systems can be eliminated by providing system air bleeds and return-line diffusers.

A 'water removal' filter or a vacuum dehydrator can be installed to remove the water from the fluid. Appendix x gives the details of water removal media.

A heat exchanger can be installed to remove the heat from the hydraulic system.

Water Removal Techniques

Moisture can be prevented from entering the reservoir in a hydraulic system through the use of adsorptive breathers or active venting systems. However, once free water is present in small quantities, water-absorbing filters or active venting systems are usually sufficient for water removal. Centrifuges, coalescers, and vacuum dehydrators are to be employed for the removal of large quantities of water. However, these methods have varying levels of water removal effectiveness. Some of the methods of water removal are explained in the following sections. Table 5.1 gives comparative data on these methods.

Adsorptive passive breather

Adsorbent medium adsorbs liquids, gases, and/or suspended matter. Activated alumina and activated carbon are conventional adsorbents used in water removal systems.

Coalescing Filters

A coalescing filter consists of a dense, inorganic fibre mat. When fluid passes through the fibres, the free water in the fluid is retained by the fibres as small water droplets. As water droplets accumulate in the fibre mat, they coalesce into larger droplets which drain down through the fibres into a collector. The accumulated water can then be periodically drained.

Vacuum dehydrator

In this method, the free water and the dissolved water can be removed by a vacuum dehydration unit with a vacuum pump, a circulation heater, and dispersion media. The vacuum pump draws the fluid into the unit through the heater, where the fluid temperature is raised approximately 65^0C [150^0F]. The fluid is then exposed to vacuum while it flows through the pores of the dispersion media. The water is boiled off from the unit, and the fluid is effectively dehydrated. This method can remove water down to 20% saturation level.

Table 5.1 | Water removal techniques

Water removal technique	Prevents humidity ingression	Removes dissolved water	Removes free water	Removes large quantities of free water	Limits of water removal
Adsorptive passive breather	Y				Prevention
Active venting system	Y	Y	Y		Removal down to <10% saturation
Water absorbing cartridge filter			Y		Removes only to 100% saturation
Centrifuge			Y	Y	Removes only to 100% saturation
Coalescer			Y	Y	Removes only to 100% saturation
Vacuum dehydrator		Y	Y	Y	Removes free, emulsified and dissolved water down to ~20% saturation

Chapter 6 | Fluid Cleanliness Standards

The cleanliness of hydraulic fluids needs to be monitored for maintaining the components of hydraulic systems at a satisfactory level. Many national and international organisations have developed standards for specifying the particle size classification and contamination concentration levels. The particle size classification standard is ISO 11171. The standard for specifying the contamination concentration levels is the most widely used ISO 4406.

Particle Size Classification Standard
The standard ISO 11171 specifies the three-dimensional size of particles, that is, 4, 6, and 14 microns, for representing the concentration levels of fine and coarse particles.

Fluid Cleanliness Level Standard
The standard ISO 4406 specifies the cleanness level of a given sample of fluid by a three-number range code representation, based on the numbers of particles of size greater than 4, 6, and 14 microns respectively, present in one millilitre of the fluid. A sample range code representation is given in Table 6.1.

Table 6.1 | Range code representation

Range code	Number of particles per ml	
	>	<=
21	10 000	20 000
20	5 000	10 000
19	2 500	5 000
18	1 300	2 500
17	640	1 300
16	320	640
15	160	320
14	80	160
13	40	80
12	20	40

For example: The numbers of particles of sizes larger than 4μ, 6μ, and 14μ are found to be 1510, 406, and 55 respectively in one ml of the sample of hydraulic fluid. Then looking at Table 6.1, the range code representation for the cleanness of the fluid can be derived as 18/16/13.

Table A5.1 in Appendix 5 gives the complete contamination code rating system as per ISO 4406: 1999.

Cleanliness Level Targets
Hydraulic equipment manufacturers, fluid suppliers, and fluid power associations have established target cleanliness levels for fluids to be used in general types of hydraulic components. The typical target cleanliness levels of fluids to be used for some hydraulic components are given in Table 6.2.

Table 6.2 | The typical target cleanliness levels of fluids

Components	Target cleanness level
Vane pumps, fixed	20/18/15
Vane pumps, variable	18/16/14
Piston pumps, fixed	19/17/15
Piston pumps, variable	18/16/14
Directional valves	20/18/15
Proportional valves	17/15/12
Servo valves	16/14/11
Pressure/Flow controls	19/17/14
Cylinders	20/18/15
Vane motors	20/18/15
Axial piston motors	19/17/14
Radial piston motors	20/18/14

Table A6.1 in Appendix 6 gives an exhaustive account of the typical target cleanliness levels of fluids to be used for hydraulic components.

Chapter 7 | Hydraulic Fluid Analysis

Fluid analysis can be carried out to ascertain the health of a fluid medium. The fluid analysis essentially counts the number of contaminant particles, detects the level of oxidation of the fluid, identifies the component wear, determines the condition of the additives, establishes the overall level of contamination, and verifies the composition of the fluid. If the analysis meets the specified cleanness target, then one only needs to maintain the filters and retest the fluid periodically. If not, appropriate corrective actions must be taken to rectify the problems.

Contamination can be measured by using portable laser particle counters. Calibrated and evaluated to the ISO standards, the portable particle counter identifies and reports the range codes for the number of particles, of sizes greater than specified sizes, present in one millilitre (ml) of the sample fluid.

A proper fluid analysis first establishes target cleanness level, sampling location, and testing frequency. Ensure that, the sample taken from the system is a representative sample of the system fluid. Fluid analysis can be conducted on a fluid sample by (1) Patch test, (2) Laboratory analysis, or (3) Online fluid monitoring, depending upon the sensitivity of the application.

Patch test
It is a simple visual analysis of the fluid sample extracted from a hydraulic system using a fluid analysis kit. It consists of a microscope, filter test patches, a vacuum pump, sample bottles, and visual correlation charts or photographs. Typically, a 100 ml measure of the sample fluid is passed through the filter media of the test patch. The patch is then dried and analysed under the microscope for both colour and content, and then compared to the reference photographs, of known particle concentration levels to determine the approximate ISO cleanness code and the type of particles captured on the patch.

Laboratory Analysis

The laboratory analysis is complete scrutiny of the fluid sample. Most of the laboratories offer the following essential fluid parameters:

- Particle counts
- Water content
- Viscosity
- Total Acid Number, TAN
- Spectrometric analysis for finding the wear metals and trending graphs

Total Acid Number is a measure of potassium hydroxide (KOH), in milligram, required for neutralising the acid present in one gram of the sample fluid.

It is indicative of the age of the fluid and can be used to decide when to replace the fluid.

Typically, it should be less than 1.4 milligrams of potassium hydroxide per gram.

A borderline value is in between 1.4 to 2.6 mg KOH/gm.

A TAN value above 2.6 mg KOH/gram is quite unsatisfactory for a hydraulic system.

Online Fluid Monitoring

With the advancement of computer technology and the introduction of sophisticated online fluid monitoring instruments, the fluid analysis can be carried out on-site in a consistent manner, while the system is in operation. Today's online contamination monitoring instruments can detect changes in the quality, contamination level, and chemical composition, of the fluid medium in a hydraulic system. It can also measure the number of wear metals present in the fluid.

Chapter 8 | Maintenance of Hydraulic fluids

An essential concern in a well-designed hydraulic system is how to maintain its fluid medium in a clean state. Remember, the fluids used in modern hydraulic systems are liable to be stressed and degraded by the presence of contaminants as well as by the severe working conditions imposed on them.

A proactive maintenance programme should be followed for the proper contamination control of hydraulic fluids. Two important proactive maintenance activities for hydraulic fluids are filtration and fluid analysis. The proactive maintenance activities in a hydraulic system ensure: (1) maximum operating performance and long-term reliability of the system, (2) reduced downtime of the system, (3) reduced overall costs, and (4) enhanced service life.

General Maintenance Activities for Hydraulic Fluids
The following activities are to be carried out on hydraulic fluids at regular intervals to keep the contamination out of the fluids:

- Clean areas around reservoirs, coolers, filters, cylinder piston-rods, fill-plugs, and dipsticks
- Clean hoses, tubes, and piping during their installation
- Check pumps, reservoirs, cylinders, and lines for leaks
- Check the fluid level in the reservoirs and keep them filled
- Filter fluids when filling
- Keep hoses capped and plugged, when they are removed or opened
- Keep the new components covered until they are ready for installation
- Change the filters regularly
- Check the appearance and smell of hydraulic fluids
- Check for the excess heat development in fluids
- Take the fluid samples regularly
- Send the fluid samples for fluid analysis

- Monitor the particulate level in fluids
- Check for the excess water content and air, in fluids
- Check for the symptoms of oxidation in fluids
- Measure the viscosities of fluids
- Replace the fluids at the appropriate time

Monitoring Hydraulic Fluids in Service
The following paragraphs describe some of the fluid monitoring activities.

Check Appearance and Smell
Very often, simply the appearance or smell of a hydraulic fluid indicates its suitability for a given application. The colour of the clean hydraulic fluid is amber. A frothy or a milky fluid indicates the presence of an excess amount of air in the fluid. A hazy or dark coloured hydraulic fluid indicates the occurrence of a high rate of oxidation in the fluid.

Some fluids have a bland, oily smell while other fluids have no smell at all. A marked change in the smell of a given volume of hydraulic fluid shows the occurrence of a chemical breakdown of the fluid. If a distinct change in the look or the smell of the fluid is detected, it is appropriate to carry out the chemical analysis of the fluid.

Monitor Particulate level
After a fluid is selected and added to a piece of hydraulic machinery, the next essential activity is the installation of correctly-sized filters at appropriate locations in the associated hydraulic circuit. This provision removes particulates and tends to achieve the specified target cleanness level in the circuit.

Check Excess Water Content

The water content in a fluid used in a hydraulic system should not exceed 100 ppm [0.01%]. The excessive amount of water in the fluid indicates the presence of an ineffective breather filter or heat exchanger in the system. The water can be eliminated from the system fluid by adsorption, absorption, centrifugation, or vacuum dehydration.

Check Oxidation in Hydraulic Fluids

The signs of the natural process of oxidation in a given volume of fluid include changes in the colour, odour, or acidity level of the fluid. Note, the presence of sludge, gum, or varnish in a system indicates that the system fluid is already oxidised. Fluid analysis can detect the level of oxidation in the system. Further, any increase in the viscosity of the fluid could be an indication of the higher level of oxidation or contamination in the system.

Check Presence of Excess Air

The ill-effects of air contamination in hydraulic fluids can be reduced by designing the system reservoir correctly, flooding the suction side of the pump with the system fluid, and providing air bleeds and return line diffusers in the system.

Check Excess Heat/Temperature

It is necessary to check the temperature of the fluid used in a hydraulic system regularly, during the system operation. If the fluid is too hot, the associated cooling system may not be working correctly, or there may be pressure-related problems in the system. Check the fluid cooler or the reservoir in the system for any failure. Remove any dirt that inhibits airflow around them. The heating of the fluid in the system may cause its breakdown.

Precautions while Handling Hydraulic Fluids

Hydraulic fluids must be handled properly to preserve their properties, avoid adverse chemical reactions, and protect fluids against the ingress of contamination. The exposure of the hydraulic fluid to the eyes may cause severe pain. Wash the eye with water when exposed to hydraulic fluid. The exposure of fluids to the skin may irritate.

The following general precautions must be observed while handling hydraulic fluids, apart from any other instructions specified by the fluid manufacturers:

- Do not keep the containers for hydraulic fluids open for more extended periods than necessary
- Wear gloves that are impervious to hydraulic fluids
- Use eye protection
- Do not expose the hydraulic fluids, especially the mineral-based fluids, to high temperatures or open flames
- Do not mix different categories of hydraulic fluids
- If the skin is exposed to the fluid, wash the exposed part with soapy water

Typical Fluid Analysis Procedure

The health of the hydraulic fluid can be ascertained by sending the fluid sample to a fluid analysis laboratory. The laboratory determines the cleanliness level of the fluid as per the relevant standards. It also determines the chemical makeup of the fluid to figure out whether the additive package is performing well as it was initially planned.

If a higher level of contamination is found in the fluid, its source should be identified to prevent its future manifestation. If the fluid is contaminated beyond an acceptable limit, it may be required to flush the system.

The fluid analysis should be carried out continuously, and sample results must be evaluated for trends that may indicate the likely change in the state of the system.

The frequency of fluid analysis has to be determined by the nature of the hydraulic application. In general, fluid sampling and testing are required for every 500 hours of operation of the system power unit or every three months, whichever comes first.

It is a maintenance technician's responsibility to collect the fluid samples from the critical locations in hydraulic systems, following the applicable standards relevant to one's region.

The following section discusses some general guidelines to be followed while carrying out the fluid sampling in hydraulic systems.

General Guidelines for Fluid Sampling

Obtaining the fluid sample from a hydraulic system involves some necessary steps to make sure that the representative sample is obtained from the system reservoir. It may be required to collect 200 to 500 ml of the fluid for complete fluid analysis.

All parts of the tiny hand-operated vacuum-assisted bottle syringe used to extract the fluid sample must be washed and rinsed with a filtered solvent, to remove any contaminants present in these parts. It is required to insert the bottle syringe to one-half of the fluid height for taking the sample fluid from the reservoir.

Remember, dirty sampling devices, and non-representative fluid samples lead to erroneous conclusions about the fluid and cost more in the long run.

The following bulleted lines give some general guidelines for taking the fluid sample from a hydraulic system:

- Operate the system for some time, say half-an-hour, before taking the sample, for ensuring uniform mixing of the fluid,
- Use an approved wide-mouth pre-cleaned sample bottle,
- Drain a sufficient quantity of the fluid from the reservoir to purge the line before collecting the sample,
- Fill the bottle with the fluid leaving a small volume, and
- Tag the sample bottle with pertinent data, including the machine number, date, fluid supplier, fluid type, and time elapsed since the last sample (if any).

Procedure for Replacing Hydraulic Fluids

Eventually, the entire fluid in the hydraulic system must be replaced with a volume of new fluid, when it is ascertained to be contaminated beyond service. Remember that the right type of replacement fluid with the correct additive combination must be used, usually as per the manufacturer's recommendation.

First, the contaminated fluid must be entirely drained from the reservoir. The reservoir should then be thoroughly flushed with a small quantity of the clean hydraulic fluid many times for making sure that the contaminated fluid has completely been removed from the reservoir. Then, fill the reservoir with the filtered new hydraulic fluid.

Each component in the system may be needed to be cleaned, and filter elements required to be changed, for removing the contamination in extreme cases.

Objective Type Questions

1. An ideal hydraulic fluid is the one which
   a. behaves like a perfect gas
   b. possesses a low bulk modulus
   c. can flow through pipes with high inertia
   d. has zero viscosity and remains perfectly stiff

2. Which is not a function of hydraulic fluids?
   a. Store energy
   b. Seal clearances
   c. Transmit power
   d. Lubricate system parts

3. What is the typical range of viscosities selected for hydraulic fluids, measured at 40 ^{0}C?
   a. 01 to 10 cSt
   b. 10 to 400 cSt
   c. 100 to 400 cSt
   d. 100 to 1000 cSt

4. Viscosity Improver (VI) can be used in hydraulic fluid to:
   a. improve the viscosity of the fluid
   b. prevent fluid breakdown against high shear stress
   c. stabilise the viscosity against the variations in temperature
   d. protect the integrity of the fluid against the harsh working condition

5. Extreme pressure additive is used in a hydraulic system fluid to:
   a. improve the load-carrying capacity of the fluid
   b. prepare the fluid for high-pressure applications
   c. prevent the metal-to-metal contact of sliding surfaces
   d. stabilise the viscosity of the fluid against high-pressure

6. Multi-grade hydraulic fluids are:
    a. fluids with VI improvers
    b. fluids without VI improvers
    c. a mixture of different grades of fluids
    d. a class of fluids with high ISO Viscosity Grades

7. An anti-wear additive added to most of the hydraulic fluids is the:
    a. Silicone oil
    b. Polyalphaolefins
    c. Polymethacrylates
    d. Zinc dithiophosphate

8. Oxidation of hydraulic fluid causes:
    a. the breakdown of fluid constituents
    b. the decrease of the lubricity of the fluid
    c. the formation of acids, sludge, and varnish
    d. All of the above

9. Demulsibility property of a hydraulic fluid allows:
    a. the heat to be quickly dissipated from the fluid
    b. the water to be readily separated from the fluid
    c. the air to be released from the fluid
    d. None of the above

10. Hydrolytic stability of hydraulic fluid indicates the resistance against:
    a. heat
    b. pressure
    c. shear stress
    d. None of the above

11. Three critical issues concerned with the proper maintenance of hydraulic fluids are:
    a. Knowing the type of contaminants
    b. Means for controlling contamination
    c. Assessing the health of the fluids
    d. All of the above

12. Mark the wrong statement
    a. Silt particles of a size corresponding to the component clearance are most dangerous to a hydraulic system than larger chip particles
    b. Fluid can dissolve water up to its saturation point
    c. The entrained air can cause cavitation and foaming
    d. An adsorbent filter can remove ferrous particles

13. The ISO standard for specifying contamination concentration levels in hydraulic fluids is the:
    a. ISO 4402
    b. ISO 4406
    c. ISO 5599
    d. ISO 1209

14. The appropriate fluid cleanness specification for a servo application as per the ISO 4406 could be:
    a. 21/18/15
    b. 20/17/14
    c. 19/16/13
    d. 17/14/11

15. For the target cleanness level of 17/15/12, as per ISO 4406, which of the following cleanness level is unacceptable
    a. 15/13/11
    b. 16/14/11
    c. 17/15/12
    d. 18/16/13

Review Questions

1. What are the primary and supporting functions of hydraulic fluids?
2. How are hydraulic fluids prepared?
3. Why are hydraulic fluids supplemented with additives during their preparation?
4. Mention five additives used in the preparation of hydraulic fluids.
5. List five properties of hydraulic fluids.
6. Briefly explain the viscometric characteristics required of hydraulic fluids?
7. What are the adverse effects of using low viscosity fluids in hydraulic systems?
8. What are the adverse effects of using highly viscous fluids in hydraulic systems?
9. Why is it significant to determine the correct viscosity of the fluid for a given hydraulic application?
10. Explain what the viscosity index of fluid means.
11. What is the primary function of VI improvers?
12. What are the effects of varying the temperature on the viscosity of the fluid in a hydraulic system?
13. Describe the lubricity property of hydraulic fluids.
14. What are extreme-pressure (EP) additives as used in hydraulic fluids?
15. What is the purpose of formulating hydraulic fluids with anti-wear additives?
16. Differentiate between the following: EP additives and anti-wear additives
17. Explain the need for excellent oxidation resistance for hydraulic fluids.
18. What are the adverse effects of oxidation in a hydraulic fluid?
19. What essential measures can be taken to protect a hydraulic fluid against oxidation?
20. Explain the process of formation and effects of the foam in hydraulic fluids.

21. What are the adverse effects of air present in hydraulic fluids?
22. What are the effects of water present in hydraulic fluids?
23. What are the adverse effects of operating hydraulic fluids at higher temperatures?
24. State the reasons for the heat build-up in the fluid medium of a hydraulic system.
25. What are the factors affecting the overall stability of hydraulic fluids? Explain.
26. What are the adverse effects of using hydraulic fluids at higher temperatures?
27. What actions may you take to improve the thermal stability of hydraulic fluids?
28. Give a brief account of the chemical stability of hydraulic fluids?
29. Give a brief account of the hydrolytic stability of hydraulic fluids?
30. Briefly explain, with examples, the compatibility issue of seal materials with hydraulic fluids.
31. Briefly explain the following hydraulic fluid property terms: (1) Pour point, (2) flash point and (3) fire point
32. What are the essential properties of hydraulic fluids? Discuss any four of them in detail.
33. What are the main classes of hydraulic fluids? List their distinguishing features.
34. Briefly explain the advantages and disadvantages of petroleum-based fluids used in hydraulic systems.
35. Briefly explain the two basic types of fire-resistant fluids?
36. What are the different types of water-based fluids used in hydraulic systems?
37. What are the essential characteristics of water-based fluids?
38. What precautions can you take while maintaining water-based fluids used in hydraulic applications?
39. What are high-water-based-fluids (HWBF)?
40. Differentiate oil-in-water emulsions and water-in-oil emulsions.
41. Give a brief account of synthetic hydraulic fluids.

42. Differentiate between the following: (1) Water-based and (2) Synthetic fire-resistant fluids

43. Give a brief note on eco-friendly hydraulic fluids.

44. What are the areas of application for biodegradable fluids? Explain.

45. How do ecologically-safe hydraulic fluids protect the environment?

46. Give a brief account of food-grade hydraulic fluids.

47. What are the general requirements for hydraulic fluids?

48. How is the fluid in a hydraulic system contaminated?

49. What are the different types of contaminants present in hydraulic fluids? Briefly explain each type.

50. Describe the effects of particle contamination in hydraulic systems.

51. Explain the effects of water present in hydraulic systems.

52. Explain the effects of oxidative reactions in hydraulic fluids.

53. Why is the cost of contamination very high in hydraulic systems? Explain.

54. What are the reasons for the fluid breakdown in hydraulic systems?

55. How does fluid contamination destroy hydraulic components?

56. What are the sources of contamination in hydraulic fluids? Briefly explain.

57. Briefly explain a typical approach to reasonable contamination control in hydraulic systems.

58. Mention some of the symptoms of excessive contamination in hydraulic systems.

59. Enlist some necessary measures to avoid contamination in hydraulic systems.

60. What are the ways to remove particle contaminants from a hydraulic system?

61. What are the ways to remove water from a hydraulic system?

62. What are the ways to remove heat from a hydraulic system?

63. State the steps you may take to avoid contaminated fluid being reused in a hydraulic system.

64. What measures may you take to maximise the life of hydraulic systems?
65. Explain the ISO cleanness standards for evaluating the contamination level of hydraulic fluids.
66. Explain the meaning of the fluid cleanness level 18/16/13 as per the ISO 4406 standard.
67. Is the ISO range code of 17/18/15 better than the range code of 17/14/11? Explain.
68. Why is it essential to conduct fluid analysis in a hydraulic system?
69. Explain the laboratory analysis of hydraulic fluids.
70. Why maintenance technician take samples of hydraulic fluid during preventive maintenance of hydraulic systems?
71. Explain the procedure to conduct the fluid analysis of the fluid in a hydraulic system.
72. What are the two essential maintenance actions in a healthy hydraulic system?
73. What are the corrective measures necessary, if the health of the fluid used in a hydraulic system is not meeting the target cleanliness levels?
74. State two important methods used for finding the fluid cleanness level of the fluid in a hydraulic system
75. What are the expected results of the fluid analysis of hydraulic fluid?
76. What are the advantages of the online monitoring of hydraulic fluids?
77. Give a brief note on each of the following: (a) Hydraulic fluid cleanness, (b) The disposal of used hydraulic fluids, and (c) The recycling and reclaiming of hydraulic fluids.

Objective type questions - answer key:
1-d, 2-a, 3-b, 4-c, 5-a, 6-a, 7-d, 8-d, 9-b, 10-d
11-d, 12-d, 13-b, 14-d, 15-d

Appendix 1

Basics of Viscosity

It is a measure of the internal resistance to flow. Thin fluids, such as water and alcohol, flow quickly, and they have low viscosity. Thick fluids, such as molasses and cold honey, pour slowly and they have high viscosity.

The viscosity of a fluid can be measured in terms of its resistive movement when subjected to an external force or gravitational force. Accordingly, there are two viewpoints of viscosity. One is the absolute viscosity (or dynamic viscosity), and the other one is the kinematic viscosity.

The absolute viscosity is the property that represents the resistive movement of different layers of a fluid when subjected to an external force. The kinematic viscosity is the property that describes the difficulty with which the fluid moves under the force of gravity.

Absolute Viscosity (μ)

A thin plate A of surface area 'a' is located at a distance 'd' from a stationary reference plate B, as shown in Figure A1.1. The plate A is subjected to a force 'F' and moves with the velocity 'v'. For small values of v and d, the velocity gradient of the particles of the fluid layers tends to be a straight line with a slope v/d.

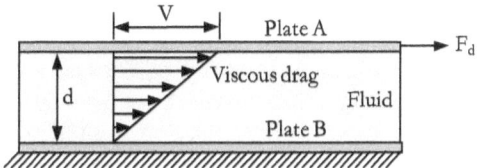

Figure A1.1 | Fluid velocity profile between two parallel plates due to viscosity

The force F is proportional to the area a and velocity v and inversely proportional to the distance d. That is,

$$F \propto a \cdot \frac{v}{d}$$

$$F = \mu \cdot a \cdot \frac{v}{d}$$

$$\text{Absolute viscosity, } \mu = \frac{(F/a)}{(v/d)} = \frac{\text{Shear stress}}{\text{Shear strain}}$$

Units of Absolute Viscosity
1 Poise = 1 dyne second per square centimetre (1 dyne.s/cm²)
1 centipoise (cP) = 0.01 Poise
1 Pascal second (Pa.s) = 1 Newton second per square metre
(1 N.s/m²)
1 Poise = 0.1 Pa.s

Kinematic viscosity (v)
Kinematic viscosity is the measure of a fluid's resistance to flow under gravity. This measure at a given temperature is given by the absolute viscosity (μ) divided by the fluid density (ϱ).

$$\text{Kinematic viscosity, } v = \frac{\mu}{\varrho}$$

Units of Kinematic Viscosity
1 Stoke = 1 cm²/s
1 Centi Stoke (cSt) = 0.01 Stoke = 1 mm²/s
SI unit is 1 m²/s

Note: Since in the CGS system, density equals specific gravity (SG). Hence, the kinematic viscosity in cSt can also be found from the following equation:

$$\text{Kinematic viscosity, } v \text{ (cSt)} = \frac{\mu \text{ (cP)}}{SG}$$

Saybolt Universal Seconds (SUS)
In addition to the basic units of measuring the kinematic viscosity, other units, such as Saybolt Universal Seconds, etc., are used for expressing the kinematic viscosities of fluids. Saybolt

Universal Seconds (SUS) is the time measured in seconds required for 60 ml of oil to flow through the calibrated orifice of a Saybolt Universal viscometer at a specified temperature. Since thick oil flows slowly, its SUS value will be higher than that for thin oil.

Viscometers
There are primarily two types of viscometers. They are the glass capillary viscometers and rotational viscometers.

Glass Capillary Viscometer
Kinematic viscosity is usually measured using the glass capillary tube viscometer with a known diameter and length. It uses the idea of measuring the time taken for a defined quantity of fluid to flow through the capillary glass tube of the viscometer. The time is then multiplied by the calibration constant of the viscometer to obtain the kinematic viscosity in Centi Stoke (cSt).

Rotational viscometer
The absolute viscosity is typically measured using a rotary viscometer. It uses the idea that the power required for turning the disk in a fluid can indicate the absolute viscosity of the fluid.

Viscosity Classification Systems
The first standard for viscosity classification was developed by the Society of Automotive Engineers (SAE) in 1911. In 1975, the International Standards Organization (ISO) in unison with American Society for Testing and Materials (ASTM) and many other standards organisations settled upon an approach to establish a viscosity measurement method. It is known as the ISO Viscosity Grade (VG), as per the ISO standard 3448:1992. ISO VG classification consists of a series of 20 different viscosity grades. Some of the Viscosity Grades are: 2, 3, 5, 7, 10, 15, 22, 32, 46, 68, 100, 150, 220, 320, 460, 680, 1000, and 1500. See Table A1.1 for more details.

Table A1.1 | Viscosity grades

ISO VG	Viscosity in cSt at 40°C		
	Mid-Point	Minimum	Maximum
2	2.2	1.98	2.42
3	3.2	2.88	3.52
5	4.6	4.14	5.06
7	6.8	6.12	7.48
10	10	9.0	11.0
15	15	13.5	16.5
22	22	19.8	24.2
32	32	28.8	35.2
46	46	41.4	50.6
68	68	61.2	74.8
100	100	90	110
150	150	135	165
220	220	198	242
320	320	288	352

The Effect of Variation in Pressure on Viscosity
An increase in the system pressure can cause an increase in the viscosity of the fluid.

The Effect of Variation in Temperature on Viscosity
The viscosity of fluids can change appreciably with a change in their temperature. Fluids have higher viscosity when they are cold and lower viscosity when they are hot. The change in viscosity with temperature is measured with an arbitrary measure called Viscosity Index (VI). Fluid having a low VI exhibits a significant change in viscosity with temperature change. A High VI fluid has relatively stable viscosity, which does not change appreciably with temperature change.

Appendix 2

Hydraulic Fluid Additives and Elements

Table A2.1 | Some hydraulic fluid additives

Additives	Elements
VI improvers	Polyalphaolefins, Polymethacrylates, and polyalkylstyrenes
EP additives	Organic sulfur-, phosphorus-, and chlorine-containing compounds
Anti-wear additives	Zinc dithiophosphate (ZDP), ashless additives
Oxidation inhibitors	Phenols, amines, and sulfides
Corrosion inhibitors	Fatty acids, sulfonates, and salts of fatty acids
Antifoam agents	Silicone oils
Demulsifiers	Ionogenic and non-ionogenic polar compounds
Pour point depressants	Polymethacrylates and condensation products

Appendix 3

Properties of Some Hydraulic Fluids

Table A3.1(a) | Mineral-based hydraulic fluid – ISO VG 32

Property	Value in metric unit		Value in English unit	
Density at 15.6°C (60°F)	0.868 *10³	kg/m³	54.2	lb/ft³
Kinematic viscosity at 40°C (104°F)	32.2	cSt	32.2	cSt
Kinematic viscosity at 100°C (212°F)	5.52	cSt	5.52	cSt
Viscosity index	108		108	
Flash point	212	°C	414	°F
Pour Point	-33	°C	-27	°F

Table A3.1(b) | Bio-degradable synthetic ester-based hydraulic fluid - ISO VG 46 (SAE 10W)

Property	Value in metric unit		Value in English unit	
Density at 60°F (15.6°C)	0.921x10³	kg/m³	57.5	lb/ft³
Kinematic viscosity at 40°C (104°F)	48.7	cSt	48.7	cSt
Kinematic viscosity at 100°C (212°F)	8.7	cSt	8.7	cSt
Viscosity index	160		160	
Flash point	220	°C	428	°F
Pour Point	-58	°C	-72	°F
Zinc	max.5	ppm	max.5	ppm

Appendix 4

Classification of fluids according to DIN 51524

Table A4.1 | Fluid classification according to DIN 51524

Type	Description	VI	Pressure (bar)	Application
HVLP	With additives that protect from corrosion, oxidation and wear	>140	>100	They are intended for universal application; however, the most significant advantage is provided when used in external hydraulic systems.
HLP	With additives that protect from corrosion, oxidation and wear	80 - 100	>100	They are intended for universal application, and they are recommended for use in internal hydraulic systems.
HL	With additives protecting from corrosion and oxidation	80 - 100	>100	They are recommended for use in low pressure internal hydraulic systems.

Appendix 5

Contamination Code Rating

Table A5.1 | Contamination code rating system as per ISO 4406: 1999

Range Code	Number of particles	
	>	<=
1	0	0.02
2	0.02	0.04
3	0.04	0.08
4	0.08	0.15
5	0.15	0.3
6	0.3	0.6
7	0.6	1.3
8	1.3	2.5
9	2.5	5
10	5	10
11	10	20
12	20	40
13	40	80
14	80	160
15	160	320
16	320	640
17	640	1,300
18	1,300	2,500
19	2,500	5,000
20	5,000	10,000
21	10,000	20,000
22	20,000	40,000
23	40,000	80,000
24	80,000	160,000
25	160,000	320,000
26	320,000	640,000
27	640,000	1,300,000
28	1,300,000	2,500,000
29	2,500,000	5,000,000
30	5,000,000	10,000,000

Appendix 6

Recommended Fluid Cleanness Levels

Table A6.1 | Typical cleanness levels, using petroleum oil, for hydraulic components. [Courtesy: Eaton Hydraulics]

Components	System Pressure Level		
	<140 bar [<2000 psi]	140–207 bar [2000–3000 psi]	>207 bar [>3000 psi]
Vane pumps, fixed	20/18/15	19/17/14	18/16/13
Vane pumps, variable	18/16/14	17/15/13	--
Piston pumps, fixed	19/17/15	18/16/14	17/15/13
Piston pumps, variable	18/16/14	17/15/13	16/14/12
Directional valves	20/18/15	20/18/15	19/17/14
Proportional valves	17/15/12	17/15/12	15/13/11
Servo valves	16/14/11	16/14/11	15/13/10
Pressure/Flow controls	19/17/14	19/17/14	19/17/14
Cylinders	20/18/15	20/18/15	20/18/15
Vane motors	20/18/15	19/17/14	18/16/13
Axial piston motors	19/17/14	18/16/13	17/15/12
Radial piston motors	20/18/14	19/17/13	18/16/13

Appendix 7

NAS Code

The National Aerospace Standard (NAS) 1638 coding system defines the maximum numbers permitted of 100 ml volume at various size intervals (for aircraft systems).

Maximum Contamination Limits (per 100 ml)						
Size range in microns						
Approximate ISO 4406 Equivalent	NAS 1638 code	5-15 μm	15-25 μm	25-50 μm	50-100 μm	> 100 μm
—	00	125	22	4	1	0
—	0	250	44	8	2	0
12/10/7	1	500	89	16	3	1
13/11/8	2	1000	178	32	6	1
14/12/9	3	2000	356	63	11	2
15/13/10	4	4000	712	126	22	4
16/14/11	5	8000	1425	253	45	8
17/15/12	6	16,000	2850	506	90	16
18/16/13	7	32,000	5700	1012	190	32
19/17/14	8	64,000	11,400	2025	360	64
20/18/15	9	128,000	22,800	4050	720	128
21/19/16	10	256,000	45,600	8100	1440	256
22/20/17	11	512,000	91,200	16,200	2880	512
23/21/18	12	1,024,000	182,400	32,400	5760	1020

References

1. Article on: 'Clean up hydraulic circuits', by Phillip Johnson, PlantServices.com
2. Article on: 'Contamination Control - A Hydraulic OEM Perspective', by R.W. Park, BE (Hons), MIE Aust., CP Eng., Managing Director, Moog Australia Pty Ltd.
3. Article on: 'Hydraulic Fluid Care Guide', MTS Systems Corporation, Minnesota, USA.
4. Article on: 'Oil Analysis 101, Part 1 of 2', by Daniel P. Walsh, Business Development Manager, National Tribology Services Inc.
5. Article on: 'Reclaiming hydraulic oil eliminates disposal problems', by W. Stofey and M. Horgan, assistant editor.
6. Article on: 'WHY OIL NEEDS ONLINE MONITORING', by Zhang Qisheng, Zhao Jingyi and Li Shuli, Fluid Transmission and Control Institute, Yanshan University, Qinhuangdao, P.R.China.
7. Article on Hydraulic contamination - part 1&2', Penton Media, Inc. & Hydraulics & Pneumatics magazine.
8. Document on: 'A guide to contamination control for hydraulic and lubrication systems Brochure: FDHB138GB1', www.parker.com
9. Document on: 'Eaton® Hydraulic Fluid Recommendations' Eaton Corporation, U. S. A.
10. Document on: 'Industrial Hydraulics', Donaldson Europe B.V.B.A., www.donaldson.com
11. Document on: 'ISO Cleanliness Levels, Fluid Service Catalog', HYDAC
12. Document on: 'Water Based and Synthetic Fire-Resistant Fluids', RA 09 296/06.98, Rexroth Bosch Group.
13. Document on Cat® Hydraulic Systems, Management Guide, Caterpillar, www.cat.com
14. Donaldson Technical Reference Guide "The Blue Pages," Donaldson Company, Inc., U. S. A., www.donaldson.com
15. Filtration Catalog Technical Catalog, Eaton, Eden Prairie, MN, USA.
16. Fluid Condition Handbook, MP FILTRI S.p.A.

Fluid Power Educational Series Books

1. Pneumatic Systems and Circuits -Basic Level (In the SI Units)
2. Industrial Pneumatics -Basic Level (In the English Units)
3. Pneumatic Systems and Circuits -Advanced Level
4. Electro-Pneumatics and Automation
5. Design of Pneumatic Systems (In the SI Units)
6. Design Concepts in Pneumatic Systems (In the English Units)
7. Maintenance, Troubleshooting, and Safety in Pneumatic Systems
8. Industrial Hydraulic Systems and Circuits -Basic Level (In the SI Units)
9. Industrial Hydraulics -Basic Level (In the English Units)
10. Hydraulic Fluids
11. Hydraulic Filters: Construction, Installation Locations, and Specifications
12. Hydraulic Power Packs (In the SI Units)
13. Power Packs in Hydraulic Systems (In the English Units)
14. Hydraulic Cylinders (In the SI Units)
15. Hydraulic Linear Actuators (In the English Units)
16. Hydraulic Motors (In the SI Units)
17. Hydraulic Rotary Actuators (In the English Units)
18. Hydraulic Accumulators and Circuits (In the SI Units)
19. Accumulators in Hydraulic Systems (In the English Units)
20. Hydraulic Pipes, Tubes, and Hoses (In the SI Units)
21. Pipes, Tubes, and Hoses in Hydraulic Systems (In the English Units)
22. Design of Industrial Hydraulic Systems (In the SI Units)
23. Design Concepts in Industrial Hydraulic Systems (In the English Units)
24. Maintenance, Troubleshooting, and Safety in Hydraulic Systems
25. Hydrostatic Transmissions (HSTs) (In the SI Units)
26. Concepts of Hydrostatic Transmissions (In the English Units)
27. Load Sensing Hydraulic Systems (In the SI Units)
28. Concepts of Load Sensing Hydraulic Systems (In the English Units)
29. Electro-hydraulic Proportional Valves
30. Electro-hydraulic Servo Valves
31. Cartridge Valves
32. Electro-hydraulic Systems and Relay Circuits

For more details, please visit: **htpps://jojibooks.com**

About the Author

Joji Parambath is a trainer in the field of Pneumatics, Hydraulics, and PLC, for over 25 years. During his career, he has trained numerous professionals from the industries as well as faculty members and students of engineering institutions.

At present, he is the key trainer at Fluidsys Training Centre, Bangalore, India, (https://fluidsys.org) which is providing training in the field of Pneumatics and Hydraulics. He has already written two books on Pneumatics and Hydraulics. The publication of the present series of 32 books is intended to restructure and update the existing books.

The author wishes to thank all trainees for their lively interaction and many useful suggestions during the training programmes that prompted the author to write the present series of books. You may send your feedback to joji.p@hotmail.com

10th June 2020